DISCOVER!

STRANGE FISH

LEAFY SEA DRAGONS ARE STRANGE

NATALIE HUMPHREY

Enslow PUBLISHING

Please visit our website, www.enslow.com. For a free color catalog of all our high-quality books, call toll free 1-800-398-2504 or fax 1-877-980-4454.

Cataloging-in-Publication Data
Names: Humphrey, Natalie.
Title: Leafy sea dragons are strange / Natalie Humphrey.
Description: New York : Enslow Publishing, 2023. | Series: Strange fish | Includes glossary and index.
Identifiers: ISBN 9781978528383 (pbk.) | ISBN 9781978528406 (library bound) | ISBN 9781978528390 (6pack) | ISBN 9781978528413 (ebook)
Subjects: LCSH: Leafy seadragon–Juvenile literature.
Classification: LCC QL638.S9 H86 2023 | DDC 597'.679–dc23

Portions of this work were originally authored by Jill Keppeler and published as *Leafy Sea Dragons*.
All new materials this edition authored by Natalie Humphrey.

Published in 2023 by
Enslow Publishing
29 E. 21st Street
New York, NY 10010

Designer: Tanya Dellaccio
Editor: Natalie Humphrey

Photo credits:Cover, p. 5 AshtonEa/Shutterstock.com; p. 4 Barbara Ash/Shutterstock.com; p. 7 Michael Warwick/Shutterstock.com; p. 9 (map) Robert Biedermann/Shutterstock.com; p. 9 (sea dragon) ymgerman/Shutterstock.com; p. 10 GOLFX/Shutterstock.com; p. 11 (pipefish) Kurit afshen/Shutterstock.com; p. 11 (sea dragon) Kjersti Joergensen/Shutterstock.com; p. 13 matthieu Gallet/Shutterstock.com; p. 15 Alexius Sutandio/Shutterstock.com; p. 16 https://upload.wikimedia.org/wikipedia/commons/2/28/Hemimysis_anomala_GLERL_4.jpg; p. 17 Thierry Eidenweil/shutterstock.com; p 19 John Lewis/AUSCAPE/Alamy Images; p. 21 Aquarius Traveller/Shutterstock.com.

Printed in the United States of America

CPSIA compliance information: Batch #CSENS23: For further information contact Enslow Publishing, New York, New York, at 1-800-398-2504.

CONTENTS

Boldface words appear in the Words to Know.

A LEAFY DRAGON

The leafy sea dragon is an amazing fish! Their bodies are covered with strange, leaf-like parts that make this fish look more like a plant than an animal. This **camouflage** is so good, even scientists have trouble spotting them in the wild.

LEAFY SEA DRAGONS CAN CHANGE COLOR TO MATCH THEIR HOME.

COZY AT HOME

Leafy sea dragons only live in **shallow** water off the south and west coasts of Australia. They prefer **coral reefs** filled with sea grass and seaweed where they can easily **vanish**. While they sometimes travel in pairs, they usually live alone.

THE LEAFY SEA DRAGON'S HOME IS SHOWN IN BLUE.
AUSTRALIA

STRANGE FAMILY

The leafy sea dragon might look strange, but it fits right into its animal family! The leafy sea dragon is in the same group as seahorses and pipefish. All fish in this family have small mouths and few fins.

SEAHORSE

THERE ARE THREE DIFFERENT SPECIES OF SEA DRAGONS.

BODY ARMOR

Leafy sea dragons have some interesting body features. Instead of scales, they are covered in bony **plates**. Their face has a long **snout** with a mouth at the end. Leafy sea dragons also have tiny, clear fins on the sides of their head and down their back!

LEAFY SEA DRAGONS CAN GROW UP TO 14 INCHES (35.6 CM) LONG.

DRIFTING IN THE OCEAN

Leafy sea dragons can't swim very fast. Their bony plates keep their body stiff while their fins help them steer through the water. This slow movement makes the leafy sea dragon look like it's drifting through the water, just like a piece of seaweed!

LEAFY SEA DRAGONS HAVE NO KNOWN PREDATORS.

NO NEED TO CHEW

Leafy sea dragons don't have teeth or a stomach. Instead, they must keep eating all the time to keep up their energy! When they find food, they use their mouth like a straw to suck it up. Then, they drift on to find more.

LEAFY SEA DRAGONS CAN STAY STILL FOR THREE DAYS STRAIGHT TO SAVE ENERGY.

A SHRIMP BUFFET

Since they can't open their mouths, the leafy sea dragon eats a lot of tiny meals. Their favorite food, the mysid shrimp, only reaches 1 inch (2.5 cm) long! Leafy sea dragons will also eat small **crustaceans** and baby fish.

MYSID SHRIMP

THE LEAFY SEA DRAGON MAY EAT UP TO 1,000 MYSID SHRIMP A DAY!

A DRAGON FAMILY

Female leafy sea dragons can lay over 250 eggs at the same time. The pink eggs stick to the tail of the male leafy sea dragon. He carries them with him everywhere for about four to six weeks. When the babies hatch, or come out of the eggs, the male leafy sea dragon leaves!

BABY LEAFY SEA DRAGONS ARE ONLY 0.8 INCH (2 CM) LONG WHEN THEY HATCH.

SAVE THE LEAFY SEA DRAGON!

The leafy sea dragon still needs our help! When the water becomes **polluted**, the coral reefs can become damaged or even die. This would mean leafy sea dragons would lose their home. Cleaning up the ocean is important to help this strange fish!

THE LEAFY SEA DRAGON HAS BEEN PROTECTED BY AUSTRALIAN LAW SINCE 1999.

WORDS TO KNOW

camouflage: Colors or shapes in animals that allow them to blend in with their surroundings.

coral reefs: An underwater hill made up of the hard parts of tiny sea animals.

crustacean: An animal with a hard shell, jointed limbs, feelers, and no backbone, such as a shrimp or crabs.

plate: One of many hard, often flat pieces on an animal's body.

polluted: Having to do with water, land, or air that is dirty or unsafe for people and animals.

predators: An animal that hunts other animals for food.

protect: To keep an animal safe from being hunted or harmed.

shallow: Not deep.

snout: An animal's nose and mouth that stick out.

species: A group of plants or animals that are all of the same kind.

vanish: To completely disapear.

FOR MORE INFORMATION

BOOKS

Murray, Julie. *Leafy Sea Dragons*. Minneapolis, Minnesota: Abdo Zoom, 2022.

Potts, Nikki. *It Disappears! Magical Animals that Hide in Plain Sight.* North Mankato, Minnesota: Capstone Press, 2018.

WEBSITES

Aquarium of the Pacific
www.aquariumofpacific.org/onlinelearningcenter/species/leafy_seadragon
Jump into more interesting leafy sea dragon facts.

Australian Museum
www.australian.museum/learn/animals/fishes/leafy-seadragon-phycodurus-eques/
Learn more fun facts, watch videos, and check out pictures of leafy sea dragons!

Publisher's note to educators and parents: Our editors have carefully reviewed these websites to ensure that they are suitable for students. Many websites change frequently, however, and we cannot guarantee that a site's future contents will continue to meet our high standards of quality and educational value. Be advised that students should be closely supervised whenever they access the internet.

INDEX